GUIDELINES FOR THE RECOGNITION AND ASSESSMENT OF PAIN IN ANIMALS

Prepared by a Working Party of the
Association of Veterinary Teachers and Research Workers

Published by
Universities Federation for Animal Welfare
8 Hamilton Close, South Mimms, Potters Bar, Herts. EN6 3QD

ISBN 0 900767 56 1

CONTENTS

PREFACE

Ever since the introduction of the Cruelty to Animals Act 1876 conditions have been laid down to regulate animal experiments including controls over the degree and duration of pain which might result. Throughout this period *pain* has never been satisfactorily defined; although it is recognised as one of the most challenging problems in medicine and biology (Melzack & Wall, 1988).

Whilst fundamental research in the physiology of pain has made great strides, little has been done to examine its assessment in animals. In recent years the need for such assessment has been recognised by scientists in different disciplines. Suggestions have been made that veterinarians, with their training and experience, should be in a good position to recognise and assess pain in animals. Veterinary expertise has also been considered to be useful for the further development of skills in anaesthesia and analgesia (Paton, 1984).

Encouragement to veterinarians to make a critical evaluation of the assessment of pain in animals led one member of the Association of Veterinary Teachers and Research Workers to issue a questionnaire to some other members of the Association. This revealed a requirement for a wider survey which in turn identified a need to prepare guidelines for the assessment of pain in animals. The present Working Party was set up to study this problem and quickly recognised that this could become a key issue in the revised legislation to replace the Cruelty to Animals Act of 1876. The Working Party has been constantly aware of the need for care in the preparation and interpretation of guidelines and would emphasise that repeated revision in the light of further research and practical experience is essential.

An initial version of these guidelines was published in the Veterinary Record (1986, *118,* 334) but since the production and publication of the first edition it became apparent that some revision and additions were necessary particularly in the introductory section and in those on the Specific Signs. Accordingly the document has been revised and, it is hoped, improved to take account of these needs. It is intended that these new guidelines will aid the recognition of pain and the assessment of its intensity and will encourage appropriate action for its alleviation.

The composition of the Working Party has remained unchanged and we acknowledge with pleasure the continued support of the Council of the Association of Veterinary Teachers and Research Workers. Valuable comments and suggestions have been received from many sources and we are particularly grateful to the following for their help:

J G Bearn, J E Cooper, G P David, C Davis, F A Harrison, A Livingston, D J Mellor,
L Murray, A S Nash, J Oldham, H Pearson, Mary Stewart, H W Symonds, Pam M Taylor, M B Tyers and others who wish to remain anonymous.

INTRODUCTION

The Animals (Scientific Procedures) Act 1986 makes provision for the control of the pain, suffering, distress or lasting harm which may result from any experiment or scientific procedure carried out on a protected animal. Applicants for project licences are asked to predict into which of three categories — mild, moderate or substantial —the procedures they propose to employ can be expected to fall (Home Office, 1986). The Secretary of State has a duty under the Act, in determining whether and on what terms to grant a project licence, to weigh the likely adverse effects on the animals concerned against the benefit likely to accrue.

In order to ensure, as far as possible, the effective and uniform operation of legislation, the need for guidelines had been recognised in a White Paper as a "significant new step in explaining to licence applicants, licensees and members of the general public the way in which the controls are expected to work" (Home Office, 1985). An important part of such guidance is information on the assessment of severity in all its forms.

Each of the terms included as contributing to the severity of a procedure requires definition and description, so that clear recognition is possible and so that the overall contribution of each to the condition of the experimental animal may be assessed. Definitions of distress and suffering include pain as a possible causative factor. For instance, the Concise Oxford Dictionary (6th Edition OUP 1976) defines distress as "Severe pressure of pain, sorrow, etc.", and suffering as "pain, loss, grief, defeat, punishment, wrong, etc." There is obviously overlap in some of these definitions. This is understandable as there is, at times, overlap in what the animals are experiencing, eg an animal in pain or discomfort may, at the same time, be suffering from emotional distress. Any attempt to write guidelines on the recognition and assessment of pain must recognise that what is being considered is not just pain but also discomfort and, at times, emotional distress.

There are widely accepted signs of health and well-being in animals. They are usually based on the appearance of the animal, its behaviour, the state of its environment and some consideration of its productivity. Animals which are suffering, ie in pain, discomfort and/or distress, are unlikely to show a full set of signs. One of the first indications that an animal may be in pain could be a change in, or absence of, certain of the signs. There can, of course, be other reasons why the signs of health and well-being are not being fully displayed. The absence of some (or all of them) alerts the observer to the possibility that something is wrong. Specific signs of pain must be recognised and assessed before it can be concluded that the animal is probably experiencing pain.

Prevention of suffering and pain

Those working with animals know from experience that avoidance or limitation of discomfort and distress is achieved by ensuring a high standard of welfare and by the reduction of apprehension in the animals for which they are responsible. Attention to welfare implies the provision of appropriate facilities and management for each species of animal kept and used for biological research. Apprehension can be minimised by careful management techniques paying attention to surroundings, the nature and frequency of human contacts, including contacts with the experimenters, and control of visual, auditory, olfactory and tactile stimuli likely to arouse fear. Avoidance of total isolation from other members of the same species has also long been recognised as effective in reducing apprehension.

Stress

Adaptive responses to changes or challenges in the animal's environment involve a variety of interlocking anatomical, physiological, biochemical, immunological and behavioural adaptation mechanisms. Combinations of these have been described as the stress response. (Some users of the term restrict it to a response centering on pituitary/adrenal cortex mechanisms, others include sympathetic/adrenal medullary mechanisms while others incorporate all the reactions described above).

Whatever the combinations of mechanisms incorporated, stress may be broadly classified into:

1. Physiological stress: the animal puts minimum effort (resources) into the response and is unconscious of this effort. Adaptation is within the normal range.
2. Overstress: the animal diverts significant effort (resources) into the adaptive response, but is again unconscious of this effort. The effort can be to the detriment of other biological processes, eg growth.
3. Distress: the animal puts substantial effort (resources) into the adaptive response. It is probably aware of the effort and may be considered to be suffering. The diversion of effort is to the detriment of other biological processes, eg growth, and the response may have detrimental side-effects.

In approaching the task of preparing what it is hoped will be practical guidelines particularly for those involved in animal experimentation, the Working Party of the Association of Veterinary Teachers and Research Workers recognised that the assessment of animal pain does present the greatest difficulties. It has therefore concentrated its efforts on the recognition and assessment of pain and the particular signs which may aid its evaluation in individual species. This document is divided into five main sections, an examination of a working definition, a discussion of the neurobiology of pain, a consideration of the systematic assessment of pain, a general scheme for its recognition and some notes on species differences.

It must be emphasised that these guidelines are intended as an initial guide to the assessment of pain and must not be taken as definitive. They will be subject to re-assessment and amendment from time to time.

WORKING DEFINITIONS OF PAIN

Pain is uncommonly difficult to define, but attempts at definition are useful in that they throw light on the process and on the nature of the difficulties encountered. Three definitions are given below:

1. Pain is an unpleasant sensory and emotional experience associated with actual or potential tissue damage or described in terms of such damage. (IASP, 1979).
2. Pain in animals is an aversive sensory experience that elicits protective motor actions, results in learned avoidance and may modify species-specific traits of behaviour, including social behaviour. (Zimmerman, 1986).
3. Pain is an aversive sensory experience caused by actual or potential injury which is accompanied by protective somatic and visceral reactions and induces changes in behaviour including social behaviour which can be specific for an individual animal.

A constructive approach to the problem of the assessment of pain in animals may be obtained if we accept that, although it is unclear what form animal experiences take and unlikely that they are as complex as those of man under similar circumstances, intact awake animals are capable of sensory experiences which (given the opportunity) they avoid and which modify their biochemistry, physiology and behaviour in species-specific ways, many of which can be related to those seen in humans experiencing pain.

There are several ways of describing pain: by means of its sites of origin, temporal characteristics and intensity.

Sites of origin of pain

Pain may originate from one or more of three main sites and be described as somatic, visceral or neurogenic.

Somatic pain originates in the skin, muscles, bones and other tissues of the body.

Visceral pain originates in the internal organs of the body including the gastro-intestinal tract and associated organs, the respiratory tract, the cardiovascular system, the urinary system and the reproductive system.

Neurogenic pain originates within the nervous system itself, arising as a disorder of processing of nociceptive activity or as a result of the generation of qualitatively or quantitatively abnormal activity in nociceptive pathways.

Temporal descriptions of pain

Acute and chronic pain are frequently used terms, but definitions of these types of pain are difficult to produce. The American Pain Society (1987) has provided descriptions for human medical practice which may be applied to animals, with some modifications.

Acute pain follows injury to the body and generally disappears when the bodily injury heals. It has a well defined temporal onset and is usually associated with objective physical signs of autonomic nervous system activity such as tachycardia, hypertension, pupillary dilation and pallor.

Acute recurrent pain is a term which can be used to describe prolonged pain, such as cancer pain, which has a definable cause and consists of repeated attacks of acute pain.

Chronic pain persists beyond the expected healing time and often cannot be ascribed to the effects of a specific injury. Chronic pain may, or may not, have a well defined temporal onset and almost by definition, has not responded to treatments directed at the aetiology of pain. The cause of chronic pain due to non-malignant diseases may be difficult to elucidate and, in contrast to acute pain, there are usually no signs of autonomic nervous system activity. Signs of chronic pain include changes in behaviour and functional ability, signs of depression, helplessness, loss of libido, weight loss and disturbed sleep patterns.

Functional dimensions of pain

Consideration of a wide spectrum of anatomical, physiological, behavioural and psychological evidence has led to the recognition of two dimensions of pain: the 'sensory-discriminative' and the 'motivational-affective'.

The 'sensory-discriminative' dimension is that which allows the subject to register the quality, intensity and site of the pain. The 'motivational-affective' dimension is that which causes the subject to register emotional and aversive experiences, to learn to avoid repetition and to carry out complex protective reactions.

Although these two aspects of pain are well recognised, the mechanisms underlying their expression cannot be easily separated one from the other and they should be considered to function in parallel, the balance between them being dictated by the origins of the nociceptive activity and the history of the subject.

NEUROBIOLOGY OF PAIN

In all animals sensory, motor and memory systems are found which protect the animal from damaging and potentially damaging interference, both immediate and longer term. Although such nociceptive systems exist in all animals, the quality of pain experiences generated by activity in the systems, and the level of awareness of such experiences, is unclear.

Pain and nociception have been best studied in mammals and man and, since we are most concerned to assess pain in mammals, this description concentrates on the neurobiology of pain and nociception in mammals and is based on recent reviews (Kitchell & Erickson, 1983; Kitchell & Johnson, 1985; Iggo, Iversen & Cervero, 1985) which should be referred to for further details.

Nociceptors and their afferent fibres

In the tissues of the body there are many thousands of receptors sensitive to a variety of physico-chemical changes in their environment. Various sub-groups signal everything from touch and joint position through to warmth and cold. Among these are groups of receptors, particularly sensitive to changes, which threaten to (or actually do) cause damage to the surrounding tissues. Thus, if the temperature of the skin rises to more than about 42°C, thermosensitive cutaneous nociceptors become active rapidly reaching their peak activity if the temperature rises a few more degrees. When such noxious temperatures last for more than a few seconds, irreversible damage can occur to parts of the skin which then die. Similarly, nociceptors in muscle are activated when the mechanical forces applied, either to or by the muscle, start to damage the muscle tissue. Substances released as a result of damage to the tissue act as potent stimuli to the nociceptors. In some organs the receptors responsible for visceral pain appear to respond to stimuli which may not be considered threatening and no pain may be registered. If the stimuli become more intense and threatening, the activity of these receptors markedly increases and pain may be registered. This suggests that specific visceral nociceptors are absent in some organs and a different method for the processing of this visceral afferent activity within the central nervous system must exist to generate pain.

There are at least three types of nociceptor: some sensitive to heating, some sensitive to high intensity mechanical forces, and others sensitive to chemicals released when tissues are damaged. Some nociceptors are sensitive to more than one of these stimuli (polymodal receptors) but this may be explained if such receptors respond to chemicals released by thermal or mechanical damage. Each group of nociceptors and their afferent nerve fibres can be recognised according to their anatomical, biochemical and physiological characteristics.

Activity of these nociceptors is transmitted to the central nervous system in two groups of nerve fibres: thin myelinated (A delta) fibres which conduct at from 3 to 30 metres/sec, and unmyelinated fibres (C) which conduct at less than 3 metres/sec. These groups of fibres provide the basis for two components of acute pain: the fast, short-lived, well localised, pricking component and slow, longer-lasting, less well localised, dull, aching component.

Activity of these receptors can be increased not only by appropriate stimulation, but also by increases in their sensitivity and reduction in their thresholds for stimulation. These changes in sensitivity can be brought about by direct action of substances on the nociceptors or by changing the inter-relationship between the nociceptor and its immediate environment. Such changes can be brought about by inflammation, by damage to tissue, by local neural reflexes or by the action of the autonomic innervation on the tissue.

Microelectrode recording from, and stimulation of, nerves in the limbs of human subjects has demonstrated that activity in each of the various types of receptor can lead to different sensory experiences, and that the intensity of the experience is related to the amount of activity generated by the receptors.

Destination of thin afferent fibres from nociceptors

The incoming fibres distribute themselves within the spinal cord or brainstem in well ordered patterns. The A delta fibres from nociceptors are distributed to the most superficial parts of the grey matter of the dorsal horn (lamina I) and to deeper laminae (eg lamina V), according to Rexed's cytoarchitectonic mapping of the spinal grey matter. The 'C' fibres are mainly distributed to lamina II (substantia gelatinosa).

Activity in these afferent fibres excites and modulates the activity of interneurones by means of a variety of neuroactive chemicals, including classical transmitter substances such as glutamate and substances with more prolonged actions such as substance P, each type of nociceptive afferent having its own set of chemical messengers and its own set of target interneurones. These interneurones, of which there are tens of thousands, act within their own part of the spinal cord or brainstem and also on other parts of the spinal cord to influence both segmental and intersegmental somatic and visceral control. Activity in nociceptive afferent fibres also directly and indirectly excites tract neurones, the activity of which is destined to reach up through the central nervous system to the brainstem and from there to the subcortical and cortical centres within which pain experiences are established. A great deal of the incoming activity from nociceptors may at times be surplus to requirements and can be suppressed by a variety of mechanisms.

Ascending pathways activated by nociceptors

In different mammalian species the ascending pathways carrying most of the nociceptive activity are located in different parts of the white matter. For example, section of the ventrolateral columns in man and monkey produces a profound contralateral analgesia, whereas this lesion in the cat produces little detectable change in response to noxious stimulation. Some degree of analgesia, however, can be produced in cats by section of parts of the dorsal columns and dorsolateral funiculus, or of the ventral columns.

Two groups of ascending pathways can be recognised, one of which is concerned more with the sensory-discriminative dimensions of pain. Those most clearly able to provide the sensory-discriminative dimensions of pain are those which are activated exclusively by stimulation of nociceptors. The activity in such specific pathways can be clearly related to that of nociceptors in intensity, spatial distribution and timing. These pathways are found particularly in the spinothalamic and spinomesencephalic tracts, though other tracts may carry such activity and in some species may be more important.

Pathways which are more likely to be concerned with the motivational-affective dimensions of pain may be activated preferentially by nociceptor stimulation but are also activated by other types of receptor. They can be activated from relatively large areas of the body, such as the skin and deep tissues of a whole limb, and can show increasing activity with successive noxious stimuli. These pathways carry activity from what are classed as multireceptive or wide dynamic range neurones, and are found particularly in the spinoreticular tract and in parts of the spinothalamic tract; other tracts may also contain ascending pathways carrying such activity in some species.

None of these ascending pathways relays all the information reaching them from nociceptors in a one to one fashion; what they do relay is a synthesis and an abstraction of information derived from nociceptive activity. In multireceptive pathways, this also includes information derived from the activity of other types of receptor. The processes of synthesis and abstraction are influenced by local systems from the higher central nervous system, which can modify transmission through ascending pathways and, under some conditions of the animal, can block transmission of nociceptive activity completely to produce analgesia.

Since complex processing occurs in the generation of activity in ascending nociceptive pathways and, since there is a large number of pathways running in parallel, it is difficult to determine even the general meaning of the activity being transmitted by a particular fibre in any pathway. Several characteristic features of pain such as referred pain, summation of noxious stimulation, analgesia and hyperalgesia can be clearly related, however, to the amount of activity and the temporal patterns in one or more of these ascending pathways. This supports the view that they participate in the mechanisms underlying pain.

Destinations of ascending nociceptive pathways

Ascending nociceptive activity acts on a wide variety of systems in the brainstem. It may reach the neurones of these systems either directly from the ascending pathways or indirectly via the cerebral cortex and sub-cortical nuclei which project back down into the brainstem.

Autonomic and somatic motor systems: Ascending nociceptive activity acts on autonomic control systems including respiratory, cardiovascular and gastrointestinal control centres lying in the caudal brainstem. The effects are seen as changes in, for example, respiratory rate, heart rate, peripheral vasoconstriction, sweating and gastric motility. Actions on major somatic motor systems, such as the tecto-spinal, reticulo-spinal and vestibulo-spinal systems, are seen as changes in muscle tone, spinal reflex sensitivity, generation of co-ordinated locomotory activity or responses such as blinking and head turning.

Descending systems modulating activity in ascending pathways: There are several descending systems which originate in the brainstem and which can modulate the generation of activity in ascending pathways, including nociceptive activity. They include systems arising in the periaqueductal-grey, raphe nuclei, lateral reticular formation and hypothalamus. They act at various levels of the ascending pathways and in particular at the level of the superficial dorsal horn grey matter. These descending systems use a variety of neuroactive chemical agents, such as opioid peptides, dopamine, noradrenaline and serotonin, to modulate the processing and onward transmission of afferent activity. Each descending pathway has its own characteristic neurochemical activity, some acting selectively on nociceptive activity to produce hypoalgesia by inhibiting generation of nociceptive activity in ascending pathways or hyperalgesia by disinhibiting the generation of such activity. These descending systems may also be activated by other pathways, some of which arise at cortical and sub-cortical level and which may not depend on nociceptive activity. It is still unclear how and when these systems operate in unrestrained normal animals although it is possible that their activity changes to limit the access and onward transmission of incoming information to that which is of immediate relevance. Techniques such as pain relief by acupuncture treatment may work by providing artificial ways of activating these systems.

Sensory-discriminative systems: These systems are the destination of nociceptive ascending pathways; they include *thalamic nuclei*. Neurones in these nuclei show activity in response to noxious stimulation which retains the details of stimulation intensity, site and duration necessary for sensory discrimination. The thalamic nuclei involved include the ventro-basal, ventro-posterior, posterior group and submedial nuclei. The distribution and numbers of neurones within these nuclei which are specifically activated by ascending nociceptive activity vary between species. The role of each thalamic nucleus is difficult to determine, but it is clear that they carry out further synthesis and abstraction of nociceptive activity in the context of the sensory and motor activity which is gathered together at this higher level of processing.

Sensory discrimination is carried further after onward transmission of ascending nociceptive activity from these thalamic nuclei to the *somatosensory cortex.* The activity is relayed to somatotopically appropriate parts of the contralateral primary somatosensory cortex and is distributed primarily to neurones in lamina IV of the cortex in these areas. Ascending nociceptive activity also reaches other parts of the cerebral cortex, including secondary somatosensory areas and widespread areas of the association cortex. Some of this activity is relayed into these areas via the primary sensory cortex, and they also receive direct inputs from thalamic nuclei which form parts of the motivational-affective systems.

Motivational-affective systems: Ascending nociceptive activity, mainly via multi-receptive/wide dynamic range neurones, reaches the *thalamic nuclei* involved in these motivational-affective systems directly, for example, via parts of the spinothalamic tract and indirectly, for example, via the spino-reticular tract. The thalamic nuclei which receive this ascending activity are found mainly within the intra-laminar group and include centralis lateralis, parafascicularis and centrum medianum. Neurones in these nuclei show multireceptive/wide dynamic range response characteristics and preferential responses to nociceptor stimulation. Differences in the patterns of activity found here compared with those of ascending tract neurones demonstrate that further processing occurs at this level. Species differences are found in the numbers of neurones in the various nuclei which respond to ascending nociceptive activity in this way.

Motivational-affective processing of pain is carried further after the onward transmission of activity from these thalamic nuclei to the *cerebral cortex* and the *limbic system.*

Widespread areas of the cerebral cortex, including somatosensory areas, association areas and parts of the limbic system, receive nociceptive inputs via intra-luminar nuclei of the thalamus. The limbic system, which includes various cortical and sub-cortical areas of the brain, is concerned with many aspects of the motivational-affective dimensions of pain. Emotional reactions and affective experiences generated within these areas are expressed in behaviour patterns such as crying out, attack or defence, co-ordinated autonomic activity and endocrine responses. Many of these responses are achieved by the limbic system acting through the hypothalamus with which it is closely linked.

SYSTEMATIC ASSESSMENT OF PAIN

Principles

1. Since a wide variety of biochemical, physiological and behavioural parameters must be considered and since adequate statistical analyses of the relative importance of each are not available for each species and type of pain, the overall assessment of welfare in individual cases can only be treated as a *value judgement based upon the experience* of those presented with the task.

   It is desirable that the experience and training of those making these judgements should be such that they are able to make use of all the relevant signs and information. It must, however, be accepted that no one person is likely to be equally well versed in all aspects, and where appropriate, advice should be sought from those with the necessary expertise. The practical experience of animal attendants can be considerable and should be fully utilised.

2. When assessment of the various parameters does not indicate the existence of unacceptable pain, then it should be assumed that there is no unacceptable pain occurring, despite the fact that such measures are only indirect approaches to the assessment of subjective experience. IF ANY DOUBT EXISTS, THE WELFARE OF THE ANIMAL MUST BE CONSIDERED PARAMOUNT AND FURTHER INFORMATION AND ADVICE SOUGHT.

3. A distinction must be made between what the animal may be feeling and what the human observer is feeling when attempting to assess pain. Ideally the assessment should not be influenced by the emotional state of the observer. Most humans have a refined capacity to appreciate or "share" the feelings of others. This capacity is not static, it appears to be learned, and its sensitivity can be either increased or decreased by experience. It is not only used in attempts to appreciate the feelings of other humans where, due to biological similarity, it can work reasonably well, but is also applied to animals when it may be misleading.

4. At least four types of response to pain can be recognised:

   a. Responses designed to modify the animal's conscious behaviour, to enable it to learn and thus avoid repetition of the situation leading to the pain being felt. These responses result from the aversive quality of the pain. High level CNS function is required for such experience and the responses elicited. Pain makes the animal feel "bad" and it attempts to avoid the situation.

b. Responses, often automatic, which are designed to protect the animal or parts of the animal. These include withdrawal from the stimulus, removal of the stimulus or attempts to achieve these results.

   A withdrawal reflex may be regarded as a minimal response and may be exaggerated through to running away. The animal may attack the source of the stimulus, eg by biting.

c. Responses, also often automatic, designed to minimise pain and which, in so doing, improve the healing process. The animal may reduce activity by lying down, standing very still or adopting some other posture.

d. Responses which are designed to convey the presence of the experience to others of the same species or other species who may help alleviate the situation, eg pain-specific vocalisation, facial expression, posture. Under normal evolutionary pressure, where predators are likely to detect such a response, it may be suppressed. In contrast, in some social species these responses are well developed and may elicit help from fellow members of the social group: this is seen particularly in primates.

These responses often occur concurrently and the "high level" subjective experiences of 4a are normally assumed to accompany those of b, c and d. The observer must be able to recognise these responses and interpret them on the basis of experience and training. Behavioural changes will often be noticed first by the animal attendant, but a veterinary surgeon experienced in this field may be best able to assess their significance.

SCHEMA

THE FOLLOWING HEADINGS AND SUPPLEMENTARY NOTES ARE DESIGNED TO PROVIDE GUIDANCE IN MAKING ASSESSMENTS OF PAIN IN ANIMALS. The emphasis will need adjustment in the light of experience and according to the species to which it is applied. Veterinary surgeons would normally carry out an examination similar to this as a routine in clinical practice.

1. Details of the animal

Should include date, species, breed, genetic strain, age, sex, weight and reproductive status and activity. Details of origin and previous experience, eg wild caught, captive bred, hand reared, etc, should also be included. Age may be a factor in the response to pain. Younger animals may be more responsive to pain, but due to a shorter learning experience, they may be less sensitive to the anxiety associated with anticipation of a procedure.

2. History of animals, establishment, etc.

a. *Signs* and course of current problem.

b. *Previous problems* in these animals or others in the same or similar environments, including both medical and behavioural problems.

c. *Stocking density,* colony or group size, cage or pen size, access to other animals including feral species and disease vectors such as flies, rodents and birds.

d. *Food and water intake* — amount, type, pattern of feeding, wastage, selection, growth, weight gain or loss, size and weight for age.

e. *Environment* — temperature, humidity, ventilation, availability of food and water, presence of toxic, infectious or allergic contaminants.

3. Clinical examination

Where an animal is suspected of suffering from clinical disease as a result of infectious, toxic or allergic agents, deficient or abnormal diets or trauma, it should be subjected to appropriate veterinary investigation and treatment. Where this state results from an experimental procedure, it will be necessary to assess whether the severity of the pain experienced is as anticipated and therefore within the conditions of the Project Licence as required by the 1986 Act. Pre-existing disease slows down recovery from experimental procedures and may mask the presence of pain.

4. Physiological signs

Most physiological signs are aspects of the integrated defence (fight or flight) reactions. These are not necessarily the same for each species. They may occur singly or in more complex combinations. Responses commonly seen include the following:

Dilatation of the pupils, wide opening of the eyelids, transient increases or decreases in blood pressure, increased heart rate, increased rate and alteration of the character of breathing (gasping or panting), movements of whiskers, piloerection, increased body temperature, increased muscle tone, sweating, changes in skin temperature, evacuation of rectum or para-anal sacs.

5. Biochemical signs

The correlation between changes in biochemical parameters and the experience of pain are as yet unclear.

Increased plasma levels of ACTH and various adreno-cortical hormones have been used in the assessment of stress; clear correlation of such increases with pain, however, remain to be demonstrated.

6. Mental status

The animal attendant is often the person best qualified to detect the presence of changes in the presumed mental state of the animal.

Mental states have been described in the following terms: dull, depressed, unresponsive, unaware, apprehensive, anxious, bright, alert, aware, excitable, hypersensitive, aggressive, timid, etc.

Assessment should be made by comparison with the previous behaviour of the animal and with the normal behaviour of the species, breed and strain. Interaction of the animal(s) with others of the same species is also important. Changes in the animal's ability to perform learned tasks, or to respond to commands previously learned, may also help to assess the mental state.

It is of primary importance to assess the *state of consciousness or awareness* of the animal by subjecting it to appropriate tests, such as response to visual or auditory threats or enticements. These responses involve the use of memory and high level perception for registration of the threat or enticement; it is a necessary condition, however, that the animal should have the motor competence to respond to such tests.

These responses are abolished by neuromuscular blocking agents and these drugs should not be administered to conscious animals.

The overall state of the central nervous system should be investigated by testing the response to a variety of reflexes, for example the palpebral reflex and flexor withdrawal reflex, to determine if CNS function is depressed and to what extent. Note that by definition, an animal under general anaesthesia cannot experience pain.

7. Abnormal activity

This is most likely to be noticed first by the animal attendants. It may range from total inactivity to maniacal hyperactivity depending on the species and the severity and location within the body of the source of the pain.

Painful conditions in particular locations may result in abnormal activity, eg pain in the mouth or teeth of ruminants leads to quidding; abdominal pain in the horse leads to kicking at the abdomen.

Other important changes in activity include changes in sleep/waking patterns, exploratory behaviour, eating and foraging behaviour.

8. Posture

Some species communicate their experiences by means of bodily attitude and other postural mechanisms. Particular changes in posture are often elicited by painful foci in particular parts of the body, eg:

a. Dogs with pain in the pharynx may stand with head and neck extended, and with abdominal pain may lie with their bellies on a cold floor;

b. Hens with abdominal pain may adopt a penguin-like stance;

c. Horses with pain in a fore-foot may point the toe.

9. Facial expression

This is an important supplement to postural change as a means of communication in some species. The various subtleties of facial expression have not been analysed from this viewpoint for most species. However, people who work with particular species often become aware of such changes without being able to describe them in detail, or to demonstrate their relationship to other means of assessing pain.

10. Gait

Pain associated with locomotion will modify the gait of the animal. This may be restricted to one limb, to the fore or hind limbs, or may involve all four.

Lameness can be graded using a scoring system. Non-painful causes of lameness should, however, be ruled out and this can often be achieved by noting the effect of the reduction of pain using local, regional or general analgesic or anti-inflammatory treatment.

Various changes in gait can be recognised for different pains in different species: eg the gait may be described as stilted, careful, etc. In many of these situations the animal is attempting to minimise the pain and by so doing will accelerate healing.

11. Reluctance to accept handling

Animals may show changes in their normal response to handling and to being manipulated. Palpation may reveal potential pain by producing changes in posture, vocalisation, attack, etc, where, up to that point, the animal had been successfully controlling its behaviour.

12. Vocalisation

This may be voluntary or involuntary. It may be "angry" or "plaintive" in tone and a variety of words have been used to describe the different noises including: growling, snarling, grunting, groaning, whimpering, whining, howling, yelping, screeching and crying. Characteristic noises may occur during a particular physiological function or behavioural act, eg grunting when the reticulum contracts in cattle with traumatic reticulitis, or yelping when a dog with a damaged bladder urinates. Noises associated with pain of more than momentary duration are usually repeated and should be differentiated from those normally made.

Systematic correlation of vocal patterns (including the use of voice prints), with other signs of pain, distress and suffering, has not yet been undertaken, although it is possible for an experienced animal attendant to develop the capacity to do this.

13. Response to analgesics

The ability of analgesic or anti-inflammatory drugs to reverse signs of pain can provide a powerful diagnostic aid. There is however a lack of adequate information on the pharmacology and efficacy of analgesics for use in many species. Where the intensity of pain is disproportionate to the injury or is serving no discernable useful purpose, effective analgesic treatment should be considered obligatory.

SPECIES-SPECIFIC SIGNS

In compiling general guidelines it has become clear that there are species-specific signs of pain which should be taken into account when making a practical assessment. Experience has taught that such signs are often associated with what is believed to be a painful condition, although no sign can by itself be regarded as diagnostic of pain and may also occur in conditions in which pain is unlikely to be a feature.

Although we have been unable to produce a comprehensive description of species-specific signs, the following notes and comments may be helpful.

General
Animals in pain often display an appearance which is best described as an "absence of normal behaviour". Normal behaviour can be characterised by various species-specific actions such as the response to stroking in cats, head pushing of goats and the general activity and vocalisation of pigs or primates. These aspects of behaviour may be strikingly absent when an animal is in pain.

Primates
Monkeys often show remarkably little reaction to surgical procedures or to traumatic injury. Obvious signs of pain are not readily seen. Loud and persistent vocalisation is unlikely to be an expression of pain, but it is more likely to signify alarm or anger.

Pain gives rise to a general appearance of misery and dejection. The animal may be huddled in a crouching posture with a "sad" facial expression and glassy eyes; it may moan, avoid its companions and may stop grooming itself. A monkey in pain may also attract increased attention from its cage mates. This can vary from social grooming to attack.

Acute abdominal pain may be shown by facial contortions, clenching of teeth, restlessness and shaking accompanied by grunts and moans.

Food and water are usually refused.

Horses
The following signs are indicative of pain; periods of restlessness; interrupted feeding with food held in the mouth uneaten; anxious appearance with dilated

pupils and glassy eyes; increased respiration and pulse rate with flared nostrils; profuse sweating and a rigid stance.

In prolonged pain, behaviour may change from restlessness to depression with head lowered.

In pain associated with skeletal damage, limbs may be held in unusual positions and there is reluctance to move, with head and neck "fixed". There may be a pain-induced tachycardia. In abdominal pain a horse may look at, bite or kick its abdomen; it may get up and lie down frequently; walk in circles; roll and injure itself as a result of these activities with bruising especially round the eyes. This state may progress and can last for several hours. When near collapse, the horse may stand very quietly rigid and unmoving but with signs of deteriorating circulatory status such as mucosal cyanosis and prolonged capillary filling time. Horses in pain generally show a reluctance to be handled.

Cattle

Cattle in pain often appear dull and depressed with head held low and with little interest in their surroundings. There is inappetance, weight loss and, in milking cows, a sudden drop in milk yield. Severe pain often results in rapid shallow respirations. On handling they may react violently or adopt a rigid posture designed to immobilise the painful region. Grunting and grinding of teeth may be heard. Acute pain may be associated with bellowing. Generally signs of abdominal pain are similar to those seen in the horse but are less marked. Rigid posture may lead to a lack of grooming due to an unwillingness to turn the neck. In acute abdominal conditions, such as intestinal strangulation, the animal adopts a characteristic stance with one hind foot placed directly in front of the other.

Localised pain may be indicated by persistent licking of an area of skin, or the animal may kick the offending area.

Sheep and goats

In general signs of pain in these species are similar to those in cattle. Changes in posture and movement are apparent and a change in facial expression may be indicative. There is a general reluctance to move. Goats are more likely than cattle to vocalise in response to pain. Grinding of teeth and grunting are also heard. It is important to note that sheep, in particular, tolerate severe injury without overt signs of pain or distress.

Following procedures such as castration and tail docking, lambs may show signs of discomfort standing up and lying down repeatedly, tail wagging with occasional bleating and with neck extension, dorsal lip curling, kicking, rolling and hyperventilation.

Pigs

Pigs in pain may show changes in gait and posture. Pigs normally squeal and attempt to escape when handled but these reactions may be accentuated when in pain. Adults may become aggressive. Squealing is also characteristic when painful areas are palpated. Handling of chronic lesions may not elicit signs of pain. Pigs will often be unwilling to move and may hide in bedding if possible.

The lack of positive signs of normal social behaviour and vocalisation may be helpful indicators of a pig in pain.

Dogs
Dogs in pain generally appear quieter and less alert with stiff body movements and an unwillingness to move. In severe pain the dog may lie still or adopt an abnormal posture to minimise its discomfort. In less severe states it may appear restless and the immediate response to acute, but low intensity pain may be an increased alertness. There may be inappetance and shivering and increased respirations with panting. Spontaneous barking is unlikely, the dog is more likely to whimper or howl, especially if unattended and may growl without apparent provocation.

A dog may bite or scratch at painful regions. When handled it may be abnormally apprehensive or aggressive.

Cats
Cats in pain are generally quiet with an apprehensive facial expression and the forehead may appear creased. There may be crying or yowling and the cat may growl and hiss if approached or made to move. There is inappetance and a tendency to hide or to separate themselves from other cats.

The posture becomes stiff and abnormal varying with the site of pain. If in the head it may be tilted or if generalised in the thorax and abdomen the cat may be crouched or hunched.

In thoracic pain the head, neck and body may be extended and in abdominal or back pain the cat may lie in lateral recumbency with back arched. If standing or walking, the back is arched and gait stilted. Incessant licking is sometimes also associated with localised pain. Pain in one limb usually results in limping or holding up the affected limb with no attempt to use it.

A cat in severe pain may show demented behaviour and make desperate attempts to escape. If a painful area is touched or palpated there may be an instant and violent reaction. There may be panting with an increased pulse rate and pupillary dilatation.

A cat in chronic pain usually has a generally miserable and ungroomed appearance with a marked change from its normal behaviour. This general lack of well-being is an important indicator of pain in this species.

Rabbits
Rabbits in pain may be apprehensive, dull, inactive and assume a "hunched" appearance. They sometimes, however, show aggressive behaviour and activity may be increased, with excessive scratching and licking. Reactions to handling

are exaggerated and acute pain may result in vocalisation. Respiratory rate may be increased and there may be inappetance.

Laboratory rodents

Pain in rodents usually results in decreased activity, piloerection and an ungroomed appearance or there may be excessive licking and scratching. They may adopt an abnormal stance with ataxia, but rats and mice may become unusually aggressive when handled and acute pain may cause vocalisation. Inappetance or a change in feeding activity may be noted and, if housed with others, a change in the normal group behaviour may be apparent.

Birds

Birds in pain may show escape reactions with vocalisation and excessive movement. Head movements increase in extent and frequency. There may be an increase in heart and respiratory rates. Prolonged pain will result in inappetance and inactivity with a drooping miserable appearance. The eyes may be partially closed, the wings are held flat against the body, and the neck is retracted. When handled the escape reaction may be replaced by a state of tonic immobility. Birds with limb pain will avoid use of the affected limb and will "guard" it from extension.

Reptiles

Acute pain in reptiles may be characterised by flinching and muscle contractions. There may be aversive movements away from the unpleasant stimulus and attempts to bite. More chronic and persistent pain may be associated with anorexia, lethargy and weight loss although it is difficult to associate any of these signs of lack of well-being specifically with pain.

Fish

It is difficult to determine the nature of the response to pain in fish and we cannot tell whether the experience is similar to that in mammals. Although they exhibit a pronounced response to injuries or to contact with irritants their response to chronic stimuli may be small or absent.

Fish with severe wounds which would cause immobility in a mammal will often appear to behave completely normally, even resuming feeding.

Fish will react to noxious stimuli such as a hypodermic needle by strong muscular movements and when exposed to a noxious environment such as a strong acid they show abnormal swimming behaviour with attempts to jump from the water, their colouring becomes darker and their opercular movements become more rapid. Such effects are indicative of some degree of distress but it is not possible to describe these unequivocally as signs of pain.

SUMMARY AND CONCLUSIONS

Despite some views to the contrary we consider that we have demonstrated that it is possible to prepare guidelines for the recognition and assessment of pain in experimental animals. The detailed information readily available to us on species-specific signs is limited and it has been particularly difficult to obtain such signs for the smaller mammals and for the non-mammalian species.

THESE GUIDELINES SHOULD BE SEEN AS THE START OF A LONGER PROCESS REQUIRING FURTHER REVISIONS AND ADDITIONS AS EXPERIENCE IS GAINED IN THEIR USE. MUCH REMAINS TO BE DONE AND WE HAVE IDENTIFIED SEVERAL AREAS FOR FURTHER ENQUIRY.

1. Collections and reviews of species-specific signs should continue to be undertaken and these should include the preparation of video recordings to be used for the training of licensees and animal technicians. Such recordings used in conjunction with the guidelines would help to establish a more uniform approach to the essentially subjective process of pain assessment.

2. Vocal patterns of each species should be analysed so that particular patterns related to the animal's experience of pain may be detected. We note that many laboratory species vocalise at audio frequencies outside the range detectable by the human ear and this must be taken into account in such studies.

3. We see merit in producing an agreed glossary of words used to describe particular responses to pain, eg those used to describe mental state, vocalisation, posture, gait, etc and to rank these, where appropriate, in order of associated severity.

4. Although it will be desirable eventually to establish scoring systems to aid in the assessment of pain (Morton and Griffiths, 1985), this will only be possible when the relative significance of all available parameters has been evaluated for each species and for a variety of types of pain. An example of how this may be achieved is provided by the system used to assess pain in young children at the Childrens' Hospital of Eastern Ontario (McGrath et al, 1985), but we consider that it is not yet possible to apply such a scale to pain in any species of animal.

In conclusion, we emphasise that, at present, judgement of the severity of pain should be based on the recognition and assessment of a wide range of parameters by trained and experienced observers. However, the less experienced observer should note that there are signs, common to all species, which may indicate the need to seek further advice from more experienced colleagues. These include changes in:

1. Behaviour patterns
2. Appearance
3. Posture
4. Gait
5. Response to handling
6. Appetite
7. Weight

It must be emphasised that none of these signs singly or together gives definite evidence of pain but, at our current level of understanding, it must be assumed that pain may be present and appropriate action taken to prevent or relieve it.

It is hoped that these guidelines will provide a framework for further assessment and action.